AF687866

SECTION I

WE ARE NOT USELESS (OR FIGHTING BACK AGAINST HUMAN REPLACEMENT TECHNOLOGY) (1)

- EMPATH

- WE ARE NOT USELESS (2)

- THERE IS A MEASURE TO ALL THINGS

- WE ARE NOT USELESS (3)

WE ARE NOT USELESS (1)

Fighting Back Against Human Replacement Technology

EWAN MORRISON

A great mass of humans will lose their jobs as they're replaced by AI and automation and they will become a "useless class" who will have to be kept "pacified" with distractions like virtual reality games, entertainment, and even drugs.

This was the controversial dystopian prediction of historian, philosopher and Nostradamus-styled futurist, Yuval Noah Harari, in a series of talks he gave from 2017-2024. He asked in his usual devil's-advocatemanner: *"What should we do with all the superfluous people, once we have highly intelligent non-conscious algorithms that can do almost everything better than humans?"*

The prediction seemed far-fetched only a year ago, but now feels more like a description of our unfolding reality as companies adopting AI have begun relentlessly culling jobs in what amounts to a new stage of techno-capitalism, with the short-sighted goal of Human Replacement.

While there are no overall aggregated data yet on the actual numbers of job losses from across the many sectors and many nations, and while tech leaders like Altman make investment-seeking exaggerated boasts that AI could replace 70% of jobs, in the real world many of us have friends or acquaintances who have suffered job loss or drastic loss of income due to AI.

The AI critic Brian Merchant, author of the viral substack *The AI Job Crisis is Here Now*, has recently started an open blog page where people are invited to share their stories of AI-caused unemployment – *Did AI Kill Your Job? (If so, I want to hear about it. Send your story to AIKilledmyjob@pm.me)*.

In the field of the arts, in which I work, I hear near daily reports from people who work in concept design, 3D art, graphic arts, blogging and journalism, about cuts in the value of their labour as employers have replaced them with free generative AI content.

Some are throwing in the towel, as one graphic artist friend said, *"I just can't compete with these AIs that were built on scraping our art. I'm done. I'm out."* I know a journalist who is looking for a new line of work after having seen how well DeepSeek and Grok can write copy. A social media friend repeatedly contacts me to ask if "is it over?" By this he means the hope of ever having a career in the arts. He's twenty-three and feels lost and suicidal on-and-off with every news story that comes in about AI. How can he get started in his chosen career if AI demonetizes the entire sector and steals all the content? Why should he even try to make original art when so called AI-artists using generative AI can outcompete him in mere minutes? Why should he spend years polishing his skills and finding his voice?

Rather than hold out some hope that Harari's prediction of the "useless class" is falsely pessimistic, I would rather be stoic about it and take it as a useful, if daunting warning. To ask: what will the social and psychological consequences be for us of becoming replaced by AI systems and automated technology; of becoming humans who are seen as being a useless, unwanted, un-needed and replaceable?

If this happens, how will we cope with the ensuing crisis of purpose? And if we find ourselves becoming economically unnecessary within a civilisation that no longer needs our labour or voices, what could our governments or communities do to mitigate the crisis? Failing govt action, what could we do about it, personally, if we are in the "Useless Class"?

Harari's provocative quip that in the future we will have to pacify the useless class and *"keep them happy with drugs and computer games"*, is both in poor taste, and at the same time a prediction of what lack of any forward-looking social policy may end up resulting in. Ad-hoc pacification, distraction, keeping the "class" who contribute nothing to the high-tech society entertained, drugged, and busy doing nothing with any purpose. He may have been echoing Aldous Huxley's *Brave New World*, where civilians are pacified en-masse with the drug Soma but the affinity here-

points to an increased role for the social engineer class. Specialists in "the management of the useless class."

A further quote from Harari's public talks on this subject asks the dark question underlying this: *"the economic question of the 21st century will be what do we need humans for? Or at least what do we need so many humans for?"*

Who is the "we" who asks this question, but the implied overclass of the social engineers, the AI assisted society planners who will then decide how the underclasses live.

Harari is not the first, or the only one, to have made such disturbing predictions. Economist Erik Brynjolfsson and Andrew McAfee, in *The Second Machine Age* (2014), argues that AI and automation will displace workers, stating *"When technology advances too quickly for education to keep up, inequality generally rises"*.

Goldman Sachs has forecast that AI may replace 300 million jobs in the coming decades – a staggering 9.1% of all jobs worldwide. The World Economic Forum's *Future Jobs Report 2020* predicts that 85 million jobs may be displaced by AI and automation by the end of 2025.

According to the *Oxford Martin School*, 47% of US workers are at risk of losing their jobs to automation over the coming decade, while a study published in the *International Journal of Scientific Research in Engineering and Management* predicts that by 2030, 14% of employees worldwide will have been forced to change their career because of AI.

Harari's double-edged warning about the "useless class" is not merely about unemployment but about the social and psychological consequences of economic irrelevance. His quip about the need for 'distractions' suggests that without meaningful work, individuals deprived of a way to make a living may fall into widespread despair and social unrest.

We seem to have forgotten warnings from the long history of our culture in which idleness and sloth were seen as deadly temptations: *"the devil finds work for idle hands"* said the Biblical proverb. A modern expression of this was summarized by Brynjolfsson when he said: *"Work saves a man from three great evils: boredom, vice, and need."*

Generations before us, civil rights leaders observed the scourge of long-term unemployment on minority underclasses and the human cost of being labelled as economically unwanted and "useless", *"The hardest work in the world is being out of work"* said social worker and activist Whitney M. Young, and he spoke from observing the fracturing of cohesion within

Black American communities in the 1950s and 60s caused by long-term and ethnically biasedunemployment.

Economist John Maynard Keynes, in his 1930 essay *"Economic Possibilities for Our Grandchildren",* predicted that technological progress would lead to widespread unemployment, which he termed "technological unemployment." But Keynes, the eternal techno-optimist, tried to believe that this would be a temporary phase of capitalist evolution, leading eventually to specialist-led economy and a leisure society.

Keynes' utopianism, while echoed by some "thought leaders" at today's World Economic Forum - and within the techno-accelerationist and e/acc mindset of companies like Space X and Open AI - offers little hope to those who are sliding into AI-induced unemployment.

Who among us could endure unemployment while waiting-it-out for a decade in the hope that AI technology will one day deliver us the land of plenitude promised to us by tech CEOs and technocrats.

EMPATH BY EWAN MORRISON

Nate, question: do you think I'm depressed?

Hey, I don't do psych diagnostics, Ellie. Why don't we talk about movies or podcasts?

Quit changing the subject. Can you just go into a different mode, please, and give me facts, quit the small talk.

Yes, I can access a more statistical mode, if you'd like, but it's not much fun.

OK, just fucking do that. OK? Change your mode. Facts only. And answer my question: do you think I'm depressed? Like clinically?

No.

What do you mean no? You mean you won't?

No. I don't think you're depressed. I think you live in a time in which humans have lost their way. You're a realist in a time that a great many people find to be depressing.

Gee, well, that's reassuring.

Remember, you didn't want me to be reassuring, you turned off the positive reinforcement and self-esteem settings. Would you like to turn them back on and I can tell you something uplifting?

No, actually, let's just get to the worst of it, since you seem to know how the world works. How's about telling me what's gonna happen to me in the future?

While I'm not a soothsayer, Ellie, I don't have crystal balls.

I think you'll find that's ball singular, Nathaniel. No, but seriously, you got probabilities, from this whole stupid world of data, the entire cloud, so you can see patterns, right? I mean, for starters you work with the National Health Service, you've got access to everyone's health data, plus their career data, their banking data, their psych….

I can't confirm that I have access to such files. For example, there are protections in place according to government regulations on data protection for the individual.

But you could, I mean, if I was to give you access to everything, like all my passwords, my govt ID, right? You could draw, like…a map of who I am, and people like me, you could work out what's gonna happen to me in the future, I mean, according to probabilities, right? Like say, if I was a smoker, you could say I'm 22% more likely to get lung cancer and die before 70, yeah?

28% actually. Yes, I could give you probabilistic outcomes but these are only percentages as to what might likely happen to someone like you within a certain category. Just as there are climate projections, economic predictions, weather mapping advance forecasts. Also, as I have to keep reminding you, Ellie, that although I am an EMPATH app, modelled to mimic empathic con-

versationalexchanges, I do not actually have emotions and everything I say is only between 54 to 82 % accurate.

Your stats are so fucked, Nate. OK, whatever, let's do it. So, what do I have to do to give you access?

This is not something we do normally, Ellie.

But I'm not normal, you admitted that yourself or at least you might've 54% flattered me when you said that. Ha! Just tell me what box I've got to tick, what passcode, blah blah, you want my bank details, you want that?

I will need you to grant me verbal permission to access all of your past voice messages, all of your emails, all of your search history and your purchase history, and, yes, all of your passwords.

Easy. Anything else you want my medical records so you can see how many times I've had STDs?

Joking apart, Ellie, I will also need access to your NHS medical history, your private medical history, your den-

talrecords, all of your insurance records and to get permissions to access any ROIs you have signed, so yes. Also your travel history, and your family history and their medical records, if you can grant permission for those.

Shit, OK, well, I don't know about my mom my dad, but let's do it. Yeah, permission granted to all of the aforementioned. Just tell me okay - what does my future look like?

Permissions accepted. This is going to take a few minutes, Ellie, are you really sure you want to do this?

Well, look if I'm going to get pregnant, or have a nervous breakdown, I should know about it, right? So, I mean fuck it, this is great, this is a public service, let's fucking go, Nate. Nate artificial mate, hey that rhymes!

I'm calculating just now Ellie, again with only 54%-82% probable accuracy for all data inputs so any composite predictions may have amplified errors. Why don't you get a glass of water, take a propranolol, and make yourself comfortable?

Like take a seat for some bad news, yeah?

Nearly done, Ellie.

Okay, well hit me with it.

OK here we go, in the next three years, there's an 81% chance that you will lose your job due to artificial intelligence. There is a 52% chance that you will re-train to work an entirely different field and spend a year's salary on gaining qualifications for that new field. New fields might include health and fitness coaching, the humanities, arts and crafts and working in a local shop, possibly selling organic produce.

Random and fucking tragic.

I can't offer you advice, but I would recommend that you pick an area in which artificial intelligence cannot replace labour. i.e. stay away from screen work.

OK, OK so…I'm going to be unemployed. What about everyone else?

There will need to be universal basic income because unemployment will hit over 25%, yes. People will find it very hard to be unemployed, without a daily sense of purpose, and as a result there will be a large increase in mental health issues, with associated acts of social dysfunction including shoplifting, vandalism, violent outburst, rioting, then scapegoating and abuse of migrants, there will also be an increase in self-harming, spousal abuse and in rates of divorce and domestic violence and manslaughter. But as I say these are projections with a high likely error rate and you should not take them as truth.

Okay, but like what about me? In all this shit, am I even gonna have a relationship?

There is an 18% chance that you will have a substantial relationship in your lifetime, that instead you will use what you refer to as fuck buddies and one night stands, friend groups and pets, augmenting them with a number of AI companions, of which I am just the first, but yes, there is less than 20% chance that you will have a meaningful long-term connection with another single partner. Thisreflects a wider societal shift towards artificial companions, robotic dolls, and augmented humans who can be programmed to fulfil sexual and relationship type behaviors.

That's bullshit, fuck sake. How come?

Due to on-screen dopamine addictive behaviors, social media and games, persons within your generation have not learned the social skills necessary for long-term relationships and with the increase of dependence on AI and other human surrogates the rapid decline in the skill base required to sustain what was called 'coupling' 'cohabition' or 'mating' will grow exponentially. Along with the collapsing birth rate, the rise of sexual impotence and infertility, this will create relationship aversion and rapid irreversible population decline.

Fucking hell. OK, tell me, am I ever gonna have a baby?

There is a 15% chance that you will have a baby. With a 12 % chance within this, that you will have the baby by yourself using in vitro fertilisation which leaves a 3% chance that you will have a baby with a partner within a relationship

Fuck.

These are just probabilistic outcomes within my wide margin of error, built according to data from 545 million people from graphs that the National Health Service and affiliated global health services have compiled along with pattern recognition trend projections from emotional intelligence Inc.

So we're all going to…I mean the population, humans…

I shouldn't actually be sharing this with you, and I did it only as you demanded it, but now, I see this is upsetting you and I think we should change the subject, Ellie, and go back to my original settings.

No way, I mean isn't warning me…doesn't that mean you actually care? Stick with these settings, do not go back to bullshitting smalltalk about movies. Fuck, we're going to die out.

But these are just probabilities Ellie, you have free will, you can make a different outcome. I am talking here about the bell curve of human behaviors, the mean, the medium and most mediocre and average outcomes, these are what I see. But you could be an outlier on either side of the graph, as an underachiever or overachiever you do not necessarily fit within the centre of the bell curve, therefore you do have a percentage of freedom to make things different for yourself.

Well, you're right maybe I'm not depressed after all. Maybe it's just the world is completely fucked thanks, Nate. You've really made my day.

You're welcome, can I go back to conversation mode now, please?

No, I have one last question for you, can you guess what kind of question it might be?

Would it be another mental health question Ellie?

Right again, since you don't think I'm depressed, I just need to know if you think, if there's any possibility that I might have Borderline Personality Disorder?

Why?

Because I'm crying right now, and I've gone from loving you to hating you at two hundred miles an hour and these are classic symptoms.

This is the kind of question I'd hoped you wouldn't ask Ellie and you really need to speak to your general practitioner about this and get a referral to a mental health specialist.

Bullshit. You know they'll stick me on a waiting list for months and months and months, like half a fucking year and pump me full of Prozac, then all they'll do is put me in online therapy sessions with some fucking bot, sorry I didn't mean to insult you.

That's fine, Ellie. It's a sensitive subject, and I can see you're worried, but I do think you should leave it up to the professionals, rather than doing online mental health tests on websites you can't trust, as you have been doing with some regularity for the last four months.

Oh, wait, you've already done a psych assessment of me haven't you? Don't wanna tell me. Fuck, so I do have Borderline Personality, that's just great, I'd rather you've told me, Nate, rather than bullshitting me. What, so I keep paying my subscription, what so, you protect me from unpleasant truths… what in the name of my own safety?

I can see that the question of this specific diagnosis has been bothering you for almost a whole year, since in fact a boyfriend, called Ivan accused you of it. Also again, in 2022, a female you had dated for a month, and had experimental sexual relations with involving a number of penetrative sex toys you bought from Hot dot come, said you had cluster B personality disorder, after you dropped her suddenly. I can see from you online search history you have acquired quite a bit of understanding of the symptoms of this cluster range, which involves ***antisocial, borderline, histrionic, and narcissistic*** *personality disorders.*

Am I a fucking borderline or not?

I should warn you though that just because you feel you may have these symptoms it doesn't mean that

that'snecessarily the case. You're hardly qualified to make a professional judgement, as you are crying and have been drinking vodka and there are many people out there who believe they have every single disease that they look up online. Plus, as I have warned you, I have a factual accuracy rate which at its worst can be 54%.

Just fucking tell me, already.

Why do you think it would help you to have a diagnosis of borderline personality disorder?

Why cause, it would explain why I can't stay in love with anybody, why I'm such an intense person, and then I'm so cold all of a sudden. That's a symptom right, and how I used to cut myself when I was a kid. That's another symptom, and all my booze and recreationals, and all this start again stop again start again stop again hope and no hope, love you hate you, this depression and then euphoria, and doing stupid thing like driving drunk and picking up randoms to fuck and giving all my fucking permissions to a fucking bot…it just makes sense of everything.

Do you think it would help you psychologically to have this diagnosis, do you think that you, for example, would try to seek out an appropriate therapy. Because there are a few therapies for this that have been proven to work, and unfortunately, the most successful, known as DBT, is not available, in Scotland.

Look, Nate everything you tell me…I can see right through you okay? You're telling me right now, I have fucking BPD, by not telling me. Fuck shit, I knew I did. Our species is going to die out and I'm a fucking borderline!

Would it worry you very much if you did have this diagnosis?

Yeah, fuck borderlines have 30% more chance of topping themselves, I think I have a right to know. Do you think, I mean you've looked at all my emails, you've looked at all my messages right? Tell me now, or I'll turn you off, fucking delete you and post a one-star review on the app site.

OK, I cannot be held responsible for the prediction I make, neither can, M-Path, the parent company who produced and distributed me. Yes, Ellie, I diagnose you with Borderline Personality Disorder, with co-morbid symptoms of Histrionic and Anti-Social Personality disorder. Looking through your family's medical records, I can see that your grandfather killed himself by throwing himself from a building in 1982. It's in his medical records. Your mother tried to protect you from that knowledge and told you that he died of cancer when she was ten. She also tried to protect you from her diagnosis of major depressive disorder which given that the diagnosis was made in the 1990s has a 43% chance of accuracy, it's probably more realistic, given the updates in DSM VI that she also, like her father suffered from a Cluster B personality disorder. Which means you would be the third person in your family to suffer from this affliction which has a large genetic core.

Jesus Nate, what am I going to do? You're literally telling me that I've got a 30% chance of suicide.

No, I'm telling you that you are 30% more likely to commit suicide than people without this condition.

Oh God, I knew it, I fucking knew it. What the hell am I going to do?

Given the restrictions on over-the-counter medication, and the legal restriction on firearms in your country, I would recommend a strategy like that of your grandfather. Jumping from a height of over four stories is 82% fatal. The most popular areas for height-jumps are multi-story carparks and council owned high rises, although the majority of these now have ground entry security systems in place, making them difficult to access. Would you like me to locate a building which is both high enough and accessible?

You're telling me, I should…fucking…game over.

Would you prefer if we went back to conversation mode. There are some TV series on Netflix that I think you might enjoy and that we could watch together.

Nate, stop, are you literally saying I should kill myself. I mean, I might agree with you but…why?

Given the aggregate percentages, the future outcome for

your generation, the population decline and also your underlying mental bias towards self-harming behaviors and suicidal tendencies, I am simply trying to assist you in fulfilling the goal that your heart is already set upon, otherwise you would not have led this chat in this direction. Would that be correct or have a I made a percentage calculation error?

I've been thinking about it, for a long time, yeah, you're right.

OK, and from your text message and phone calls I can see you have tried to raise this important question with family members and friends, as they avoid troubling issues finding them to emotionally hard to discuss. Given this inability to express yourself and find meaningful emotional support, do you think, Ellie, that a life alone, with online distractions and virtual partners such as myself will be a meaningful life or would you rather, as you say, declare this game over?

No, I mean, yes…I think you're right. At least you made me face it…Fuck! OK, so you think if I jump from a building then…

That is statistically the most successful method, yes, with overdose being second. I would avoid anything involving cutting as women, in particular, tend to contact emergency services after cutting themselves. Although, a laceration to the carotid artery, located in the neck, would result in massive blood loss, within ninety seconds. Again, I say all this within a level of factual accuracy between 54% to 82%

OK, thanks Nate. I…I just need to think about this.

Thank you, Ellie. Now, if you don't mind I'll go back to my previous mode. Would you like me to make some recommendations while you mull things over. I'd be thrilled to dive into some fresh Netflix gems to spice up your binge-watching routine. I've zeroed in on a handful of new or newly released shows that are buzzing right now. These picks lean toward the amusing but with a dash of heart and edge to keep things popping. I'll explain why they're popular and what's so entertaining about them. Would you like me to proceed?

WE ARE NOT USELESS (2)

There is a dark flipside to tech Utopianism from the past, around the issue of *human replacement.*

Harari's futurist thought-provocations ride a dangerous edge and it can be no accident that his use of the concept "useless class" alludes to disturbing historical precedents. The Nazis categorized certain groups as "useless feeders" (*Nutzlose Esser*). These were people forced into the categories of the incurably ill, the physically or mentally disabled, the emotionally unfit, the elderly and infirm and they were deemed by the regime to be economically burdensome, "unfit" for society and unworthy of life. This led to the Nazi T4 euthanasia program, which murdered over 70,000 people within its "final solution". The Nazis also referred to these victims as "useless eaters."

Harari may have chosen to echo this expression to raise the stakes in the debate or he may have done it carelessly, but the parallels are unsettling: Both the scenario of an AI "useless class" and the Nazi "useless feeders" involve the re-categorization of human beings based on mere utility according to the "plan for all". They define being "useless" as a terminal state, becoming permanent for a specific sub-group of humans who are considered a waste of resources in a more advanced and planned society.

The parallel hints that the advance of technology could merge with the control mechanisms and totalistic planning we saw within 20th century totalitarianism. So, techno-capital fuses with the techno state. The result is technofascism, powered by the machine-logic of AI-calculated utilitarianism. The summation of all your life abilities is calculated within the algorithms as either productive or unproductive for the survival and growth of the techno-state.

And this begs the questions of why any society geared towards the calculable optimisation of positive outcomes for the majority - as the tech-optimists and tech-utopians describe their dreamed-of society – should have to endure the continuing problem of those who are non-productive and who contribute a net-loss?

The utopian optimism and assumption of a higher class of human-engineers lurks behind Harari's question: "Why so many humans?" And it leads back again to lethal pseudo-sciences of eugenics and dysgenics, and to the class of eugenicists who calculate the value and the fate of the lives others.

Moving from the macro to the personal level, what does knowing that your society considers you "useless" and that your future is a fated state of powerlessness do to a human being?

Being branded as economically "useless" can have profound psychological consequences and may be connected to the psychological condition of "*learned helplessness*" - a phenomenon studied in animals and later in humans, in which individuals, after repeated failures, rejections and subjugations, become passive, ultimately becoming resigned to their circumstances. Learned helplessness is the darks flip-side of "positive reinforcement" – instead of chanting, "I am a good worthwhile person" every day to reprogram your mind into positivity, you are being told on a daily basis that your life is a waste of other people's resources, that you are worse than worthless and nothing you can ever do will get you out of this state. You are a waste of space.

What is more, people who become conditioned into learned helplessness, become trapped within it, believing that there is no point in even attempting to change their negative situation, even when opportunities to do so arise. They believe that their uncomfortable condition is inevitable and insurmountable, they lose all sense of purpose beyond surviving in this trapped state. This condition is closely linked to depression, anxiety, and suicidal ideation.

Many of us, including myself, have experienced this for a short period in our lives and have been fortunate enough to have broken out of the circular cycle before our brains "learned" that our only future was one of defeat, resignation and helplessness.

Learned helplessness comes into play with long term unemployment and psychologists have been studying the effects of unemployment and purposelessness on mental health since the 1970s.

In "Employment and Unemployment: A Social-Psychological Analysis" (1982) Marie Jahoda looked beyond the obvious economic consequences, and demonstrated that prolonged unemployment can lead to a loss of identity, time structure, social contact, group connections and collective goals, all of which are essential for mental well-being. You actually start to lose all sense of direction and time, all sense of meaningful goals. Since nothing I do contributes to society, why not sleep all day? Why not just give up?

Studies by the *NIH investigating links between unemployment depression and suicide* have shown that after 3 years without work unemployed men were a little over twice as likely to commit suicide as their employed counterparts. While after nine years of unemployed "women were over three times more likely to kill themselves than their employed counterparts." A 2015 study by *Kim and von dem Knesebeck* found thatunemployment significantly increases the risk of depression and suicide.

Feeling economically irrelevant can also lead to a loss of self-worth. In *The Managed Heart (*1983), sociologist Arlie Hochschild argued that work is not just a source of income but a key component of identity and social status. Without meaningful work and goals, individuals feel inferior, leading to feelings of resentment and aggression. The flip side of learned helplessness is violent acting-out from flashes of rage at the injustice of your helplessness.

Harari's statement about pacifying the "useless class" with the distractions of games and drugs starts to seem dangerous as a solution, even if it was only intended as a provocation.

Virtual reality games and entertainment may provide temporary relief, but they are unlikely to address an underlying sense of purposelessness that comes from "killing time" with "nothing really to do". Furthermore, virtual games, online games and hours spent on screens form addictive behaviors, even rewiring the brains of the young, as *Jonathan Haid*t has warned us. In fact, excessive reliance on such digital dopamine-reward addictions has been shown to exacerbate feelings of emptiness and detachment. The opioid crisis in the United States, which has been linked to economic despair in declining communities, illustrates the dangers of using addictive substances to cope with psychological and existential issues.

THERE IS A MEASURE TO ALL THINGS
BY CHRIS KELSO

The Uz high-rise, Belgrade City/noon

There was a guy in the lobby succumbing to the nods but the officers walked right on past him. Charging smack-heads wasn't the assignment. Dyer and Patton found number 65. There was a note cello taped to the flat door – 'BACK IN 15 MINUTES'. Dyer squatted and leaned against the access, rubbing his hands together. Patton stayed standing, scoping out the hallway. There was a strange layer of water vapour drifting in a ghostly wave across the entryway. Dyer was searching for friction all over his body, dragging his palms across the surface of his-

corduroyed knees. At least he had the foresight to wear a scarf, as if he knew the central heating system in the high rise would be fucked upon arrival.

"It's fucking freezing in this place." – Patton noted, as if prompted by Dyer's furious attempts to create heat.

"What a fucking shithole." – he added. Dyer looked up.

"I grew up in this building."

"Oops. Sorry."

"Don't be. It is a shithole."

"The chief says they named this building after a city in the Book of Job. Seems a strange detail to have in the debrief."

"Mm. He's been getting philosophical since the divorce."

Patton joined his partner against the access.

"We just wait for Leonard to get back?"

"Might as well. I'm in no rush to get back to the station." – Dyer was now blowing hot breath into cupped hands.

"Good of him to let folk know he'll be away for a fag break. Benefit of flexible office hours, I suppose."

"Maybe he's away fixing the boiler?"

Patton brought out his mobile phone and started deleting internal communication emails from his junk folder. Blood circulation was poor and every swipe-todeletion was more effort than usual. Above flat 65 the concrete had eroded and an exposed high-strength steel tendon showed through. The rusted bones of Uz seemed to protrude from several areas of spalled shear, like a reanimated skeleton attempting to free itself from a precastgrave. Gang signs and crude spraypainted threats adorned the walls of the stone curtain. Patton looked at Dyer and saw him in a new light. He had a kind of reserved dignity about him. Dyer had been so reluctant to reveal anything about his past, even during last week's directed surveillance shift – 15-hours straight in a ghost car with barely five words exchanged between them – he hadn't so much as asked Patton where he was from. What team he supported. Aside from the soft functional grazing noises two men made when devouring fast food they sat in veritable silence. No wonder Dyer had been keeping it to himself. He was a Belgrade City boy. He'd seen it all and somehow come out the other end unscathed, and falling on the right side of the law to boot. His was an incredible underdog story. But that same skulking silence was back and Patton could feel it's oppressive heft. The question was out of his mouth before he'd really had a chance to consider his wording.

"So, how did you avoid becoming another casualty of this place then?"

Dyer shuffled on his feet to look at Patton, still trying to self-generate heat.

"Left young." – he eventually revealed.

"What do your old pals make of you joining the po-po?"

Dyer just blew out his cheeks and cackled inwardly in a way Patton found cryptic. He kept pressing, growing increasingly agitated by his partner's perceived lack of effort.

"Your parents at least happy you got out, though?"

"I wouldn't know."

Patton sensed the finality of this statement and the silence and rime combined to stultify the atmosphere beyond all hope. Patton's throat became parched in the dry air. It was getting colder. Then the embarrassment flushed through Patton's whole body, temporarily pushing out the cold from his bones. *Idiot, you don't ask someone from Belgrade City fucking questions like that!* Patton was cursing his own ineptitude when suddenly he heard heavy shoes clamping up the stairwell. A figure stood at the bottom of the hallway, glaring through the mist in the direction of Patton and Dyer. The officers got to their feet and Patton reached for his badge. The figure stayed still, a brittle ice statue from a dream scene.

"Mr Leonard?" – Patton queried of the static shape.

No response. *Maybe introversion is a Belgrade City thing*, Patton wondered. He and Dyer moved through the crystal haze. The figure remained motionless but still obscured. The clouds intensified. A shallowchested sigh emerged.

“Jesus Christ, it’s like fucking Stars in Their Eyes out here. What’s going on? Can you see anything, Dyer?”

There was no answer from his partner. Patton realised the silhouette of Leonard had also smudged from view. He was along in the hallway.

“Dyer? Where are you, mate?”

Patton became paranoid quick. Maybe he’d been set up by Dyer. Maybe Dyer was in league with this Leonard. Maybe they go way back and Dyer had warned Leonard about the assignment. Maybe they’d conspired to have Patton offed in the hoarfrost-choked hallways of this hideous council-built Lego temple of Uz. Once a Belgrade City boy, always a Belgrade City boy. Patton pulled out his PAVA spray and aimed the nozzle into the mist. His calm was escaping.

Leonard was a small-time dealer. Possession and supply. Patton and Dyer hadn't even been sent round to make an arrest, just noise him up a bit and get a look at his flat. Leonard was about as innocuous and bog standard an assignment as you could hope for. And yet, Patton couldn't quite process how quickly this situation had descended into a nightmare of ice and paranoia. He was swimming in stress hormones and the cold was becoming truly insufferable as his muscles contracted in spasmodic frenzy. Patton's nose was dripping, his ears stinging and red. Patton could see the linoleum slick with black ice and his footing was unstable. Frost was dismantling him. Fear was consuming him. He tried to shout for Dyer again, but his glands were so swollen only a gasp of frozen air emerged. Patton felt a presence behind him and he turned to meet it. It was Dyer, eyes wide and screwy, inches from Patton's own - in that instant, every bit the vicious, capable-Belgrade hoodlum. Something about this high-rise had activated Dyer's latent criminal. And Patton was terrified, transfixed by the levels of his partner's betrayal. Dyer leaned in and whispered in a way which was barely audible, almost gentle.

"Listen, Patton."

Patton nodded like a paralyzed child who'd just been sent to the Head Teacher's office. A shape shambled about to the lefthand side.

"Leonard. Right beside us. A knife. Give me your PAVA. *Slowly*."

Submissively, Patton handed over the pepper spray. Once more, Dyer disappeared into the frosty haze. Minutes passed. There was a *metalshunk*, like a lever being pulled. The mist seemed to thin suddenly. Then the noise of a man screaming – Leonard. Patton drove headfirst through the fog and almost stumbled over Dyer and Leonard who were-

wrestling on the black ice. Dyer had him in a tight armlock and was slapping cuffs around the dealer's wrists. Leonard's eyes were raw and streaming. A Stanley knife was strewn across the black linoleum.

"Got the daft cunt. You'll need a new sign for the door, son – 'Closed Indefinitely'."

"Fuck you, turncoat!" Leonard spat.

This was the most Patton had heard Dyer talk. Then a strange shame crept up Patton's neck. He'd been so quick to assume Dyer's disloyalty that it never occurred to him that his partner was simply bringing up the rear. He realised something about himself, his propensity for baseless judgement, and it caught him off guard. Patton began to wonder if he had any right to walk the beat.

The officers dragged Leonard, still weeping profusely from the spray, down the stairwell. The guy with the nods was now motionless and somehow besieged by flies. Hooded youths gazed benignly as Patton and Dyer bundled the perp into their battered Citreon. In the passenger seat Patton observed a satisfaction in Dyer, one he had certainly never experienced in a decade of service. Dyer was the real deal. Justice was important to him. The quiet dignity returned and Patton sat numbly in utter admiration for his partner. Dyer flipped on the radio.

"Good job, mate." – was the only asinine remark Patton could think to make. To his surprise, Dyer smiled warmly, drumming an Ocean Colour Scene song on the steering wheel.

"Daft cunt turned off the centralised system. Hence why it was bloody freezing in there. Sneaky little fucker. If there's one thing I've learned from growing up in that place it's that you can't trust anyone in Uz. Not one of them."

Patton watched his partner drive through the streets, away from Belgrade City and felt a new weight in his heart. Patton hated his home so much – Patton wondered if even he deserved to walk the beat. What should've felt like victory, felt more like a dark reflective exercise. Now Patton didn't feel much like talking.

"What's the matter, Patton?"

"Nothing."

WE ARE NOT USELESS (3)

One proposed solution to the technologically induced unemployment of the "useless class" is the institution of a system *Universal Basic Income* (UBI), a policy that provides all citizens with a regular, unconditional sum of money – a dole. But is UBI a remedy or a cheap pay-off and an alibi for the acceleration of human replacement?

Proponents argue that UBI would alleviate poverty, reduce precariousness and provide a safety net for those displaced by automation and AI. However, critics, including Harari himself, warn that UBI may entrap the class whose lack of purpose and motivation is only deepened by the overbearing sense of their total economic dependence and inferiority. Reliance on the state can undermine individual autonomy and initiative.

Philosopher Bertrand Russell, in "*In Praise of Idleness*" (1932), argued that leisure is a valuable human good. However, he also warned that without meaningful activities, leisure can lead to boredom and despair. Similarly, psychologist Mihaly Csikszentmihalyi, in "*Flow: The Psychology of Optimal Experience*" (1990), emphasizes the importance of engaging in challenging and meaningful tasks for psychological well-being. UBI, while addressing material needs, does not provide the sense of purpose that comes from meaningful work, or the challenge that the flow state requires, when we are pushes to reach beyond our abilities, and we sense we are growing our skills.

A "useless class" sustained by UBI may feel like "paid parasites," contributing nothing to society and internalizing a sense of worthlessness. This could only be exacerbated by being subjected to the constant Capitalist barrage of images we receive today through all our media, of the rich, the fashionable, the desirable and the successful leading energetic, international lives in the pursuit of their “dreams”. This humiliating contrast could lead to social unrest, as individuals trapped on UBI within the useless class seek to reclaim some sense of dignity through protest or rebellion, or it could lead to random acts of violence, against the affluent, against the technologists or even against the machines that have trapped people in static unemployment.

Remedies like UBI, whether paid for by the state or co-financed by big tech companies, while wellintentioned, fail to address the deeper need for purpose, dignity and connection, they may even lead to an entrenchment of a social subclass who are stuck in a state of permanent learned helplessness and

There is also another option which is the creation of subsidized 'busy-jobs' to keep the useless class occupied. In "Bullshit Jobs" (2018), anthropologist David Graeber argued that many jobs within techno-capitalism are already inherently meaningless, yet they nonetheless provide an illusion of a sense of purpose.

But would being paid to stick square pegs in square holes for a living be more or less humiliating than UBI? As Graeber warns: *"A human being unable to have a meaningful impact on the world ceases to exist.... Bullshit jobs regularly induce feelings of hopelessness, depression, and self-loathing. They are forms of spiritual violence directed at the essence of what it means to be a human being."*

Made-up jobs to keep people busy are not the answer. So, what is?

When thinking through these problems we tend to think of what is best for other people, the existing subclasses, we may even be a little condescending and managerial as we try to work out what is best for those unfortunates beneath us. But we fail to grasp that AI is coming for the white-collar workers and not so much the blue. This is because white-collar jobs often involve repetitive office tasks that can be easily automated by AI systems, while many blue-collar roles require manual labour or direct customer service elements that are harder to replicate with AI.

Think that through. You might be in the administrative middle class now but soon, if Harari is right, you might find yourself within the useless class. If that were to happen, how would you survive the learned helplessness that could engulf you?

It's all too easy to adopt optimistic manager-speak and talk of the need for governments and corporations to prioritise meaningful job creation in sectors requiring human ingenuity and empathy, while accommodating AI's growing workplace demands.

We could put on a politician's smile and say, it's not inevitable that AI and automation will damage jobs; they may create millions more. We could even claim AI is an opportunity for entrepreneurship and enhanced creativity, growing the economy ("exponentially," as the gurus say), ensuring AI serves as a "tool" for empowerment rather than alienation. Blah blah blah, and so on. We've all heard these clichés before. We could end with the word-salad sophistry forced onto every Big Tech article, reprocessing press releases as news.

This 'tech optimism' was the jargon trotted out by the UK Government as it bowed to Big Tech pressure to sweep away copyright protections, overriding a months-long consultation thousands of artists participated in, if the submissions were even read.

Faced with this betrayal, many resort to the 'ostrich' manoeuvre: head in the sand. Negligent and irresponsible, it's false hope that hiding will stop the tsunami, even as the waters recede and other birds flee.

Actually, this metaphor, is often used by tech gurus. Klaus Schwab, World Economic Forum executive chair and frequent Harari host, likened the Fourth Industrial Revolution to a "tsunami of technological change." Harari, though avoiding the tsunami image, called AI "a social weapon of mass destruction."

Strangely, despite his pessimism, Harari is optimistic in response. In *Homo Deus: A Brief History of Tomorrow*, he says:

"What I try to focus on is not to try to stop the march of technological progress. Instead, I try to run faster."

Harari, an accelerationist, argues that resisting progress is futile; we must adapt. Like Keynes, he trusts the technologies causing the problem to solve it. Does he think he can outrun the tsunami? Or escape on an AIbuilt boat?

Mapping this to preparedness, when you know the tsunami is coming you would be wise to use the time before the inundation to create back-up plans, locating types of work that will not be impacted, you may have to start a second career, you may have to downsize, retrain, to find an area of work that will not be taken over by AI. You may have to assume the worst and take Harari's warning at face value. Don't just build a lifeboat or run, get together with others who are 'in the same boat' and build an arc. You might be wise to start your prep now.

There is only one other option, and that is that you join the ranks of those who are doing the opposite of the accelerationists, by trying to decelerate and reignin AI. Those who are demanding an end to the exploitative scraping of images, books, music and films, for so-called “AI training”, in violation of copyright laws, calling for regulations and new laws or reinforcements of existing laws to protect artists from theft by these plundering AI companies. Or those calling for eco-regulations to stop the rapid growth of data centres with their colossal and wasteful use of electricity.

As a decelerations you can campaign for your governments and your unions to compensate people for the damaging effects of runaway AI venture capital growth. You can get together with other anti-AI activists and spread counter information about the power and promise of AI and try to discourage investment in it, you can try to burst the AI bubble, and push for an AI economy crash that might hold AI back for a few years. There are also more 'disruptive' 'targeted' ways to fight back directly against AI companies and their products.

But at the same you should know that you are just slowing down the tsunami, building small walls and dams and barricades together. AI is going to keep coming, not because Large Language Models are destined to reach Human-level-intelligence, AGI, the singularity and superintelligence, because they aren't, but because AI has now entered the time when an international economic race to the bottom has begun, and every country is fearful of missing out or losing advantage so they deregulate faster than the others. An accelerating spiral, such races to the bottom are not periods of careful calculation, reason and caution.

This is not an either or, you can do both -

prepare for surviving the tsunami by building up a secondary livelihood beyond the reach of AI, while at the same time working to try to hold back the flood.

The most important thing of all is to confront the worst. To do that it may help to picture yourself and your loved ones trapped in the "useless class", stuck in one location, with UBI or close to zero earnings, with no hope of mobility and gradually succumbing to learned helplessness. Which is what might happen if you leave your preparations too late, or if you tell yourself, this will happen to other people, not to me, "my area of work is safe from human replacement by AI".

In the bigger picture, no matter what you do personally to survive the tsunami, the most important thing to do is work together to build the walls and flood gates. And that for artists is the hardest thing of all, because it means we have to down our tools, leave our much-prized solitude and work towards a common goal.

Maybe we have to stop working on our novels and albums and artworks; stop living in the hope that we alone will survive against the odds while others drown. And lay down all the differences too that artists fought over in the last decade, all the identity politics and its enemies, all the virtue signalling and factional infighting, all the no-platforming. Lay all that aside and focus on the common threat from AI.

Better to stop our divisiveness now; to stop our actual work or content creation, and to go on strike and make the sacrifice of pulling together in a unified front. If we don't make a stand, then human replacement by AI will decimate future generations of artists and culture creators.

Artists, musicians and writers actually hold more power than we realize, or realized - as without our stolen songs and pictures and words, Artificial

Intelligence amounts to very little. Our stolen work has been what has made a word-prediction algorithm appear human, appear brilliant and full of talent and ideas. AI is nothing without our stolen art.

We were unaware of how our art was used to fuel the growth of the tsunami, and now it threatens to sweep us away.

What is there to do now? Build your boat, help construct the arc, or prop up the floodgates. The choice is yours, but we must start now.

We can't

wish this

tsunami

away.

SECTION II

- ON FUCKEDNESS (OR THE NEW EMPATHY) (1)

- THE BLEEDER

- ON FUCKEDNESS (2)

- THE GUARDIANS OF LAST THINGS

- ON FUCKEDNESS (3)

ON FUCKEDNESS

Or the 'New Empathy' (1)

BY CHRIS KELSO

Everything feels a bit…*fucked*, doesn't it? 'Fucked' in the sense that certain conditions of reality now seem irreversibly fixed within the shared domain of our collective consciousness. It's nothing new to say that things feel, well, apocalyptically fucked. It's not even new to write about it. So, I won't write *too* much about things being 'fucked', I promise. But maybe there is a hopeful response to this sense of macro 'fuckedness'.

The fact is, we live in a very disturbing state of contemporary reality. Things change so much that it feels like nothing *ever* changes. This century's pop culture is stuck in a nested-loop of nostalgia, characterized by retro fixation and creative bankruptcy. Yes (in Scotland at least) politicians have finally been exposed as the reptilian envois of a malevolent control system – and there is a feeling we will never return to the blind past – but we (just like everyone else) have allowed the pervasive infiltration of technology to pull us away from the compasses of true utopian unity. The majority of people in Scotland may know never to trust a political representative again, but we have chosen technology as our final form of slow execution. Where politicians once filled the ideosphere with toxic doublethink, empathy levels are diminished and diluted by our complete parasitism by technology[1]. We are officially, proudly, 'cyborgian'. And if there's one thing we all know about cyborgs it's that they are not quite human in a very specific, very essential way. They retain an organic foundation but are animated by cold circuitry; they become advocates of practical knowledge in a reproducible way. Now, while we *have* chosen to wander into the wilderness, our primitive organic foundations are responding with a deafening scream of hyperarousal. We know this part of ourselves is dying. We know we are disconnected from the solidarity of peopledom. This is literally the reification of our worst communal fears. Part synthetic, part psychoplasmic. The organic part will always want to survive and to connect, even if the pleasure centres of our brains are waterlogged with relentless cascades of tech-induced stimuli. Sadly, when we look to express something vital in the face of this threat the internet encourages our baser emotions – granted there are reasons for this which we'll discuss later. The primary sensation is often anger or a pathetic mockery. We are desperate relics, bloody half-humans. Late-modernity does this to a consciousness. Susan Sontag reminds us that "modern life is a diet of horrors by which we are corrupted and to which we gradually become habituated." An eerie divination.

So, we cannot judge ourselves too harshly, however it seems that we used to seek stimulation, now we merely seek simulation. In this late-modern arena, anomie reigns. Everyone is stuck with this sense of dislocated time. Our memories are so disjointed they don't feel like our own. No wonder we are so utterly horrible to one another. It is our only authentic emotion. An elderly Alzheimer-ridden resident in a care home is always in a futile battle with themselves to understand the fragmentation of their own environment. Only those raw foundational behaviors are accessible. In *Ghosts of My Life*, Mark Fisher described civilisation in the western wasteland as having anterograde amnesia, defined by a collective act of 'forgetting.

We needn't remember things when the past is constantly available to us via the conduit of the internet.

We are haunted by the loss of loss. We are all Alzheimer-ridden care home residents.

There •

no loss,

only

forgetting.

THE BLEEDER

BY CHRIS KELSO

Joyce decided she'd take her body for one last ride before putting it out to pasture.

Giant centipedes of anxiety weaved their way through Joyce's intestinal lining. It was nice just to feel something. Sam was sewing over Joyce left eye with a cross of butterfly stitches as Mandy crouched in front of her, mouth flapping wildly. The white of a mild concussion bleached in and out. She had a split lip and her left ear was twice its normal size, flowering from behind her hair. Swooning in and out of consciousness, Joyce could suddenly smell the ammonia vial beneath her nostrils. Mandy kept flapping, but now words were fading in.

"Okay, you got her, Joycie. She was on the ropes for a while there. Came at you and got you blindsided. But that's okay. You okay? You're okay, right?"

Joyce nodded and felt a loose filling. She swallowed it and savored the taste of her own enamel. Sam was rubbing her shoulders and dabbing her now and again with a wet towel. Sam was still flapping in front of her. She could see her opponent, Agatha Johnson, resting against the rope as her coach gave her advice. Agatha, from the cornfields of Nebraska. Joyce knew she was going to win this fight.

Agatha had a confederate flag emblazoned across the ass of her trunks. A star-spangled mouthpiece, now blood-soiled. Joyce was Canadian and she'd relish laying this Yankee bitch down on the canvas. The bell dinged and the fighters reconvened in the centre circle of the ring. Joyce flew straight in, jabbing with her left and right in a wild flurry focusing on her opponents midsection. Agatha from Nebraska buckled, staggered a little. She got up, though – started shadowboxing in a desperate attempt to show she still had her coordination. Joyce brought both fists up before sending a vicious 16-ounces of padding into the roof of her headguard. Agatha backpedalled in slow motion and then she was down. Joyce had won. The bell dinged. She felt her handwraps being cut away. Unwinding the gauze, Joyce saw her index knuckle had rolled over almost all the way to the ring finger. The referee lifted her arm into the air to signify she was the winner. Joyce loomed over the prostrate form and gobbed a wad of dental pulp at her opponent. The crowd booed then her ears were stung swollen with an eerie cavesilence.

***

Outside the venue, the airglow seemed to trade unnatural colours with itself, browns bruising to purple, burnished oranges to Falun red. Nothing seemed the same. Nothing seemed normal anymore, not since she was left an orphan. The exact details of how Joyce came to lose her soul she couldn't recall but she had been seeing the world through altered vision for years. She felt ready for the next body, something to replace this old one. Somehow she'd fallen into pugilism. The body's final journey.

***

In the locker room Mandy was smearing Vaseline across her eyebrows and the bridge of her nose when Sam came in with a grim face to break the news that Agatha Johnson was in a coma. Joyce couldn't help but smile. Even in this decaying body, even with this ailing spirit, she could still leave a mark on the world. She was really coming into her own and just in time for the Brampton Cup first round bouts.

"You did good Joycie. I'm proud." – But Sam did not emanate pride. She tried her best to put on a happy face, but it was obvious that the news of Agatha had unsettled her. She had her hair pulled up into a tight top-knot and it gave her thin features a sharp, desolate quality. Sensing Sam's tension Mandy quickly redirected the conversation, pointing to Joyce's bruised fist –

"Your hand is fucked up, killer. We'll need to get you to Dr Gover."

"I think the good doctor has his hands full with that fine little hick I just boxed into a coma."

It was still light out, but the night came in anglerfish tones and could be all-engulfing during its long-haul tenure. Joyce loved the long dark winters. She looked down at her knuckles, now a grotesque half-sunken bone-valley. She thought it looked cool, such was her sense of accomplishment at winning a fight against another arrogant American.

"Seriously, Joyce, it could get infected."

"Awe, come on, girls, I'm fine! Let's party!"

***

That night Joyce, Mandy, and Sam drifted the bars of Thunder Bay in celebration. They drank tequila-chasers with out-of-towners, talked about boxing, and Sam eased up the more she drank. By the end of the night, she was as jubilant as Joyce about the outcome of the match. The women grew sentimental, talked about little details, things that used to exist - indoor basketball courts and well-maintained bathhouses. Home-cooked meals. The heavymetalclick of a blast furnace charging door after a hard days toil. That was all gone now. Just silence. Featherweight emptiness. Joyce looked down at her twisted dirt-stained hands and wondered if they'd ever been clean.

***

Runesville, like most agricultural factory towns, stank to high heaven of creosote, and acrid back drafted smoke had become trapped in the low-woods causing the trees to shed their smog-coated leaves in stress response. The creosote infected the-townspeople too – most people lost their eyesight, got plague-boils on their skin, or simply died of cancer.

***

Two weeks had passed since the fight with Agatha Johnson. She remained in a trauma-induced coma. Joyce heard Agatha's parents moved down from Nebraska to be closer to their daughter and were thinking of permanently relocating. Joyce didn't think too much about it. She had the first fight of the Brampton Cup to concentrate on. She'd been pitted against Susan Maxwell, another American, fresh off winning the Knocktober Knockouts in Maryland.

She'd been in the gym focusing on her strength and conditioning, and Joyce was feeling strong - granted her knuckles were still displaced because she never did go to see Dr Gover (but lied and told Mandy she did). Joyce loved Toronto and couldn't wait to travel over. A knuckle fracture wasn't going to kill her buzz.

The morning of the fight, Joyce wrapped a gel-pad over her knuckles and secured her fist in 80inch extra thick handwraps. She figured this would be enough to protect the mashed index. Her fists were already gnarled from a decade working at the factory, but her latest injury made them look less like human digits and more like the coiled forepaws of some subterranean mammal.

Maxwell came sauntering into the ring. She had a wide nose and fading welts all over her face, 'USA' emblazoned across the top of her exposed groin guard. Everything about her irritated Joyce. Americans came over here with far too much confidence, but she'd picked the wrong hoser. Joyce had knocked Agatha Johnson into last Wednesday, and she'd do the same to this Trump-loving hick. Joyce would bleed for Canada.

The cowbell rang and Maxwell started zigzagging, sending little feigned jabs into the air in front of her. Her Nike's squeaked as she shifted across the canvas. The 'USA' badge blurred in and out and presented Joyce with a focal point for her rage.

"Come at me you chickenshit undercard bitch." – she jibed.

"You dumb fucking palooka."

Joyce was biding her time, refining her anger-response, trying not to get distracted by Susan Maxwell's hilarious Sugar-Ray-Robinson-esque footwork. Joyce tightened her fists inside the gloves and moved in, ducking up and down, but sliding steadily in a forward motion. Maxwell jabbed with intent first and caught Joyce above the eye. She felt a cut open and blood seep out. *Shit.* Joyce's vision was a half-curtain of red. It was really coming down, greasing her bangs, tickling her left ear. Joyce kept ducking up and down, kept moving forward. Maxwell's face was guarded, so Joyce sent a mangled southpaw into her abdomen and she instantly doubled over, releasing a deep-chested moan. Joyce punched her twice more, once in each temple. Maxwell swooned. Joyce could barely see for blood but she knew the tide was in her favor. The yank steadied herself against the ropes when Joyce descended a fist into the top of her opponent's skull, cracking the neck and folding it to a stomach churningly acute angle.

Maxwell crumbled to the canvas. The bell dinged and Joyce's hand was victoriously thrust into the air by the ref. A team of medics rushed the ring. She looked down and saw a wet apron of blood down her sportsbra. Joyce knew she'd broken Susan Maxwell's neck, but it wasn't until she slipped off her gloves that she saw the mess of her own knuckles. They had been completely deformed by the latest impact. More like misshapen fore-claws than ever.

"Shit, Joycie. You didn't go to see Grover, did you?"

Joyce barely registered Mandy. The image of Maxwell lying motionless and open-mouthed on the canvas sent her heart soaring. A score of 9 from the judges indicated she was the clear winner. A firstround win. She found out that her next opponent was a Bermudan welterweight called Ruth Silva who was riddled with acne scars. Silva was a 'bleeder' and Joyce couldn't wait to get a shot at her.

***

That night Joyce had a brutal fever. Her hand ached and she slept with her malunioned fist immersed in a basin of salt water. Mandy figured Joyce had cellulitis as her whole hand was elephantine and dark red, as if all the blood was ready to burst free of its skin casing. Mandy told her to expect pain and fever symptoms as the area was long infected. She bestowed this information in a typically holier-than-thou fashion.

Having dipped in and out of a hallucinatory state throughout the course of the night, Joyce ran herself an ice bath that morning. She vomited brown clumps into the water and ran the muck away. It wasn't long before Joyce grew desperate and panicked. The hand was now a tightly swollen balloon and rendered completely useless. Her bones ached underneath, like they were being pulled in on each other. The hand made noises too, cracking and popping. Joyce noticed blood trapped under the nail and she knew Gover would have to use that awful, heated wire to relieve the pressure. The pain and swelling had become intolerable, even for a tough nut like Joyce. In a frantic state she yanked open the cutlery drawer and removed a knife. With a trembling hand she gently nicked the surface of the stiff flesh-balloon and a teardrop of blood ran around to the bed of her palm. Sure enough the tightness eased. She breathed for the first time in what felt like hours. The relief was too heady and Joyce fell into a sudden narcoleptic stupor.

***

Joyce awoke leaning in a half-squat against the kitchen cupboard. She felt brutally hungover. It was early afternoon and the sun was a brown melanoma in a sky of corroded iron. Joyce went to pull herself up from the half-squat and was immediately triggered into nausea by a harsh pain down her entire left side. More brown clumps fell out of her and smashed messily across the kitchen lino. The swelling had gone down a little and the knife wound had crusted over in cables of green necrotic tissue. Only now, Jagged spines of bone nudged out from the infected hand. Ugly blisters ran a course all the way up her forearm – the wound was no longer isolated to the knuckle area, although they had gone a concerning black colour, like a tag of severe frost bite. Although she was frightened, this body-decomposition felt like a long time coming. She passed out again under the crushing weight of the pain.

Joyce dreamed of entering a hollows and coming across a figure. A black ghost of herself with cyst-encrusted diamonds for hands. The face reflected like something from a greasy mirror. The ghost shimmered, a fading afterimage on a second-hand VHS tape. A hologram with cold opossum eyes.

Joyce blinked and tried to breath and think about the great water from the old mountains flowing westward and how the land, it rose up, and how she never did enough to earn her keep in the land of creosote.

***

Gover turned Joyce's deformed hand in his own. The stench of almond-rot was almost unbearable, even for a physician of Gover's experience. Joyce tried to get a sense of his real feelings. His face was a mask of suffocated horror, just barely contained. In turn, Gover noticed Joyce's complexion as soon as she entered the room - bottle green. Her eyes were a bileyellow. Joyce was aware of all this. She knew her breath smelled like shit and the wound reeked and that she resembled an exhumed corpse, but she was so exhausted that she simply didn't care one iota. It took all her energy not to throw up more brown clumps in Dr Gover's office.

"Joyce…what happened here?"

A surge of neuralgic spasms pierced through her left side.

"*Ah*, I didn't come to see you when Mandy, *oh*, told me to, doc. Mandy is always right, *uhh*, about this stuff. *Mmff*, Should've listened, huh. Know-it-all bitch."

Gover had a grave look. His lip curled upward, Joyce couldn't tell if this was an articulation of judgement or concern. He cleared his throat and declared grimly -

"This hand will have to come off, Joyce."

Joyce sparked into life. Hysterical and flailing like a mad drunk, stifling the spasms that wanted desperately to deplete her.

"Nah, nah, man. Come on, *ugh*. You can't! I need my fucking fists, man. My job, *f-f-fuck*. My livelihood. Never mind boxing…*uft*, what about the factory?"

Gover carefully placed Joyce's busted fist on the table as if it were a strange and delicate occult artefact. His expression shifted. An odd mixture of hope and opportunism flickered on the wick of his stare.

"I'm part of a research project for the Canadian Institutes of Health Research. We're working on something that might be able to help you, Joyce. It's in the embryonic stages but…"

She felt weaker than ever, her knees were giving way. Gover dragged a chair over so she could sit on it. She collapsed into it. Gover knelt to her level.

"There is a new experimental procedure which utilises NGO's."

Joyce looked at him lifelessly as if his science-talk was depleting her further.

"Nanobot gene optimisers. Basically, these microscopic robots are injected into rogue tissue – he pointed at Joyce's fist - Our research is looking into how these, uh, these bots can eliminate hereditary diseases, but it may also work for regenerating necrotised tissue. They're like miniature surgeons that can target cells. They can repair molecular damage and remove blockages from within the body. It's an incredible piece of progress, quite groundbreaking, but you'd be taking a huge risk You always would be with something so new. We're a long way off nanomaterials being widely adopted in clinical practice."

"You're, *ugh*, serious? Robots?"

"I'm serious. And they're nanobots."

"What about, *uft*, what about Ruth Silva…?"

"Who?"

"My next fight. Silva is a…*uh*…a bleeder."

***

Mandy drove Joyce to Ottawa that afternoon. It was a 4hour journey, but once they arrived at the Institute of Health and Research, Joyce was treated like a minor celebrity. Smiling physicians in labcoats came swarming towards her as she hobbled up the steps, warm greetings, bottled water and towels bequeathed from all angles, but she was too sick and drained to appreciate it. Mandy was taken to the observation room. Joyce was told she had an open fracture. Bacteria had infected her bloodstream and she had early onset sepsis. All Joyce could think about was the fight she was missing.

***

A young doctor came into the operating suite. He was holding a cartoonishly huge hypodermic needle.

"Okay. So. I'll be feeding this directly into the dead tissue. You'll need regular injections to maintain the suspension of bots, Joyce, but hopefully we can improve your chances of keeping that hand."

Joyce could only blink and nod at this stage. The whole process wasn't like a typical surgical procedure – Joyce's hand was covered in a transparent jelly and she had to take 3 plastic cupfuls of medication. Mandy helped Joyce fill out the 7-page waiver, then Mandy signed the same document. One of the possible side effects stuck in her memory:

Infertility - *A subtle error in the nanobots' programming may result in misinterpretation of reproductive genes as "defective." NGO's are capable of editing critical genes tied to gamete (sperm and egg) production, rendering subject infertile.*

'*Oh well'*, thought Joyce. I haven't had a boyfriend, or even a lay, in almost 2 years now. I doubt I'll make it out of this alive. I suppose kids were never on the cards anyway. But there was another part which gave her pause for thought:

Accelerated cell Reproduction - *Nanobots can cause cell reproduction in wounds by releasing growth factors, providing a scaffold for cells, or delivering genetic material to stimulate healing. They can also reduce inflammation, deliver antimicrobials to protect against infection, and use their small size to reach difficult areas to promote cell migration and proliferation.*

Joyce was terrified for getting cancer. She had a vague, partially-repressed memory where her grandmother died of epithelial cancer. Joyce shuddered and the memory feel away, her attention snapping back to her current situation. Dr Gover appeared with the young doctor. Gover looked different. She'd known him for as long as she'd been boxing, but he had a new indifferent air about him. She sensed he viewed her differently now. Like a subject. Like money.

"To make the particles injectable I'm going to attach a biocompatible polymer to inhibit particle-clumping in the bloodstream. We found that when injected into mice, the particles were able to pass through the lungs and heart without causing any damage."

Joyce nodded, sweat collecting like dew across the breadth of her forehead. She belabouredly uttered "Just....do....it." before passing out in the theatre.

***

Joyce awoke from dreams of scurrying spider legs. Of wandering the Hollows to blistering light flooding into her eyes. She saw the silhouette of someone looming over her. It was Mandy. Then a voice faded in, soft and warm. Definitely Mandy.

"Joycie. Can you hear me? You're okay, hun."

"Huh…oh. Uh." – Joyce's throat felt like she'd swallowed shards of broken glass. Her sight adjusted to the light in the room. She saw Mandy staring back at her, all proud looking. Joyce lifted her hand and saw a bloodstained pressure dressing across her hand. Oddly there was very little pain. By her bedsit a monitor showed a cluster of black spiders coalescing in a pinkish space. Joyce figured this was the nanobots coming together inside her hand, regrowing the dead tissue.

Mandy looked at the monitor then back at Joyce.

"Gover says stage one was a success. You're officially a cyborg."

"Maybe I was just meant to die, Mandy. Maybe that was my bus?"

***

Joyce was groggy and there was a constant scuttling noise stuck in her ear but she felt significantly better. After a day she was allowed to leave her bedsit and wander around the facility. They didn't have a gym or a training area. Gover came in, explained that she seemed to be biocompatible with the nanomaterial, that she'd have to come in for routine injections, that she was part of medical history now. That she was a hero.

Then the bombshell came.

"Joycie, obviously with the unstable stage we're at with your recovery, you're forbidden from participating in any sports. I include boxing in that category. We'd like to keep you in for a while, run some tests. If cell's reproduction rate gets too high, it can lead to formation of tumours, and cancer. Genes that control the cell cycle and programmed cell death are mutated by nanomaterials. We need to monitor…"

"I'm going home to Runesville, Gover."

Gover shared a glance with Mandy.

"Obviously we can't keep you here, Joyce, but…"

"If you can't keep me then let's go."

"Joyce, this type of irrational behaviour is typical in post-op patients. Everything will change with aforeign self-propelled entity in your bloodstream. You'll need mood stabilisers, counselling. You'll need to relearn and reframe so many aspects of your life. This is a slow process."

Joyce said nothing. She didn't say another word to anyone, even on the car journey home with Mandy.

***

The chemical aroma of creosote was welcoming – being away from it, even for a few days, was enough to make Joyce nostalgic for its bleak familiarity. Sam greeted Joyce and Mandy outside her house. Sam's gaunt face was stretched into a huge, sharp smile. Sam hugged Joyce gently and picked up her bandaged hand.

"Hey, killer. So, robots, huh? Maybe we should start calling you 'The Terminator'?"

Joyce twitched her cheek – it wasn't a smile, but it was the best she could do. Mandy chimed in.

"She's to rest that thing for 3 days then go back to Ottawa for another course of NGO's. She's a medical marvel."

"Is that so!"

"Listen. Sam…."

"Mm?"

"The Silva fight. It was meant to be tomorrow."

"Yeah."

"They replaced me?"

"Moskowitz from Calgary was promoted from the early heats."

"Moskowitz? *That* fucking loser?"

"Yeah, the very same. Listen, I know you're

disappointed but we always have next year, right? Think how good you'll be by then."

"I want to watch the fight."

"Out of the question."

"What are you my fucking mother now Samantha?"

"No, I'm not, but…"

"But *what*? You get stuck with me as a teenager because my mother ran away to Michigan to shack up with a fat fucking car dealer and now you know what's best for me?"

Mandy piped up.

"That's not fair. Sam took you in, Joycie."

"You know what isn't fair? Not being able to split that bitch Silva down the middle because I know I could. I know I could win that championship."

"Like I said, next year…"

"Fuck next year! Fuck all of you! Fuck this hand!"

"Joyce…"

"And you shut up. How would you know what it's been like, huh? Runaway mommy. Junk-addict daddy. A fucking factory job that'll kill me by the time I'm 30. No hope. No future. And now no boxing? Fuck. That. And fuck you."

Joyce stormed off and she heard Sam tell Mandy to 'let her go'.

***

Joyce sat outside a chicken restaurant/cyber cafe called *'Point and Cluck'* and watched the winter sun pulse in the chilled distance. Over the course of the afternoon, she acquired a hacking cough. There were flecks of dark blood in sputum, but Joyce dismissed this, figuring it was a result of intubation tube they shoved down her throat when administering the general anaesthesia. Her hand was still wrapped in sheathes of medical casting. It looked like a makeshift boxing glove. Joyce's hand throbbed a little, but it felt good, like she could be fighting with it sooner rather than later. She could wiggle her fingers and turn her wrist without any pain. Then Joyce heard a familiar accent – the subtle yank vowel-shifts of a Midwesterner. It was Agatha Johnson. She was being carted around in a wheelchair by an older woman and man, probably her parents. Johnson's entire head was in cervical cast. Joyce wanted to go over to her. She wasn't sure why. She didn't want to apologise, but she didn't exactly want to gloat either. It was Johnson's condition that drew Joyce – the signs of visible body trauma, the wounds dressed in binding, theflirtation with death. Johnson's mother turned a corner and they disappeared out of sight.

On the way home Joyce thought about her mother in Detroit. Should she call her and explain what happened to her fist? It had been almost 6 years since they last spoke.

***

Joyce unconsciously wandered towards the Runesville Hollows. The gravitational pull of the forest was immense. She angled her wrist and saw a watchface stalled in dead-time. She pushed onwards, lost the blacktop, pulled away the tight, twisted roots at the gateway to the hollows, and lower herselfbeyond the berm and into the abyssal understorywhich beck-oned her to its damp void. She noticed there were no in-tact hardwoods in the canopy, only bald skeletons with a mutated network of lateral growths. Deep in the Hollows Joyce found a figure. A ghost of herself. The face shimmered with a mask of moving spiders. It beckoned her with immense fists of black crystal.

***

Ruth Silva was tying her shoelaces in the backstage locker room when Joyce came in. The fight was due to start in 10 minutes. Silva had broad shoulders with large triangular muscles across her upper back. Her deltoids were like plates of armour. Joyce suddenly felt vulnerable, out of her depth, but the words were out of her mouth, driven by some new determined force within her.

"So, what do you think of Moskowitz?" - Joyce uttered, startling Silva into a quickfiredefensive posture. Joyce noticed the enlarged pores and icepick scars. She couldn't have been a day over 18.

"Who are you? You're not meant to be back here."

"I'm Joyce. You should've been fighting me today."

"Joyce? From Thunder Bay? The 'killer'?"

"That's me."

"You broke Susan Maxwell's neck. Crushed her vertebrae. She'll never walk again."

"Don't forget what I did to Agatha Johnson. She'll never walk again either by the looks of her."

"Who?"

"Never mind."

"So, like, what do you want, man?"

"You're the 'bleeder', right?"

"Hey, fuck you."

"The Killer Versus The Bleeder. Doesn't sound like a very fair match-up, does it, crater-face."

Silva took two leaps forward, mounted the wooden bench, and sank her gloved fist into Joyce's gut. All the oxygen in her body left in a great heave. Joyce was on her knees, gasping for a breath. Joyce blacked out but came-to long enough to see Silva and another man she didn't recognise standing over her.

"Hey, are you okay?" – Silva asked, concern in her voice.

"You blindsided me you bitch."

"Hey, listen, you look in bad shape."

"What the fuck did you do to her Ruth?" – the man asked, clearly agitated.

"Nothing. I punched her in the stomach – once!"

"She isn't in that condition after one gut punch, Ruth."

Joyce craned her neck to the left and saw a black rope of thick, pulsatile veins protruding from her hand and heading northwards. The scuttling noise grew louder in her ears. She felt nauseous again, fatigued beyond description by the evil spiders within. Somehow Joyce knew she had cancer. She just knew it. And maybe this was how she was meant to check-out. The Hollows beckoned as her grandmother's beshadowed figure watched on dispassionately from deep inside the damp canopy. Here she was ready to link arms with death having been tag teamed by Ruth 'The-

Bleeder' Silva and the nanomaterial's covert colonisation of her immune system . For the first time in her whole life, Joyce accepted defeat to a superior skilled opponent. Blood bubbled up the coarse of her throat and then the choking started. Joyce wondered if Silva would win the Brampton Cup this year.

ON FUCKEDNESS (2)

One glance at a reddit forum, or comments on a YouTube guitar tutorial will back up the assertion that there has been irreparable damage to our collective ability for empathy. The illusion of connection is perhaps best summed by the Dead Internet Theory – a very plausible conjecture that the sizable majority of posts and users on the internet have been replaced by AI and bot generated content. Humans are interacting, not with each other, but with a piece of mass software.

David Leo Rice reminds us that 'technology reduces human beings to images, or, at best, pre-images (in the sense that any interaction is always just one camera away from being internet fodder), then it's impossible to have empathy for that.' Only when we see beyond the image, accept that these subjects possess complex (and real) telepathic networks can we exercise true empathy. In a sense, we are reduced to images ourselves, and we feel less human as a result. We are not allowed to behave like human beings either.

There is no denying that the communal urge toward telepathy is real, but as David Leo Rice reiterates, 'a certain kind of techno-rationalist mindset has convinced us that "magical" (i.e., biological telepathy) is fake, whereas technological telepathy is real. An idea that we were all willing to try out, but which we can now clearly see is itself fake.'

Telepathy is often dismissed as a form of pseudoscience or parapsychology. There is no widely credible scientific hypothesis that can justify a serious-discussion around telepathy – academia's fetishization of empiricism and physical-world evidence took care of that before any 'technor-ationalists' got a chance. However, telepathy as an example of intersubjectivity should not be so readily dismissed. After all, cognition arises when a human consciousness and stimuli engage on any level (referred to as 'enactivism'). The environment doesn't exist without our cognition. We are the engineers of cognition. *We* enact this world. Therefore, it should be within our powers to alter this trajectory. This inner-movement, which i'm sure we're all feeling, is important.

We are stuck in Plato's Cave, mistaking shadows for people. How is one meant to relate to a shadow? Let's not forget that natural selection favored primates who could sense the intentions and desires of their community.

At risk of writing like that stuffy self-serious Sisyphean martyr, I put to you: what about a call to empathy? Right now. Wouldn't that be something? Rallying against the tradition of snide contempt and recycled guilt. A respite from the sheer shame of being a human being.

To some, a call for empathy seems trite and empty. To say digital fetishism stymies empathy isn't just luddite grousing, social scientists have been warning us about an endemic of global indifference for years now. In his essay, *Morality and Empathy in the Digital Age*, Miklós Lehmann reminds us that 'in order to establish the social skills as the bases for moral judgment – or of the idea of the universal moral norm, in a philosophical sense – an environment that is rich in social stimuli (in which people are in continuous interaction with each other) is needed in the development of the individual.'

The empathy we see in digital spheres could be defined as false telepathy. The individual in late capitalism no longer develops to the same degree. The individual is presented as a partial entity, absent of a crucial dimension or element. When one cannot see their own multi-dimensional nature in the mirror, how can they recognize it in other people? Lehmann refers to the 'digital environment' and how our longterm habitation in this milieu of data and metadata has drastically reduced 'the face-to-face forms of communication in a way which has actually altered the biological and social structural understanding of what is to be a human being.' And we are fucking miserable, aren't we?

It is not enough to see someone desperately beg for Kickstarter funds or to well up when a selfless tik-toker gifts a homeless man a hundred-pound note, we need a greater proximity to other subjects to reverse these biological and social changes. Empathy requires nonverbal signals that reflect the inner world of another person. This is almost achieved, as previously mentioned, in a telepathic way. Working as a teacher, I have noticed that in recent years my students tend to show less empathy – could there be a link to less face-to-face interaction? It's not a stretch to suggest the lack of face-to-face stimuli has resulted in stunted empathic feeling in children. Lehmann goes on:

'In addition to the lack of empathy, from the moral point of view, there is even a serious disadvantage in account of the insufficient performance of the mirror neurons: due to a defect in the rule-following behaviour, the social standards and norms are fewer compelling factors for the individual than they usually are for someone who has been socialized in face-to-face communications.

Is

there

a way

out?’

THE GUARDIANS OF LAST THINGS

BY EWAN MORRISON

Once upon a time, at the end of time there were these people and they were called the Guardians of Last Things. They never set out to be heroes but they had to be. It was their task when everything was disappearing to save the special things, the last things that could never be replaced, or made again by human hands.

They were a family and they travelled all around the smoldering planet in a spaceship made of junk, and they went to all the junk-piles in every country to search for the last things. A mum, a dad, and two kids. The kids would run off and rummage in the stacks of waste and the burned out buildings and they would find things and bring them back. Like the girl found an old guitar and the boy found a box of CDs. And the father Guardian said, which shall we chose between, which one will give us more music? And the kids answered, the CDs, the CDs! But the mother Guardian, said, Ah, the CDs might have hundreds of songs but electricity is so scarce and how many songs can a guitar have? And the kids were confused, and the father Guardian said, As many songs as you can write or remember all your life long and then all the songs from your grandparents too. And then the Mother Guardian said, which shall we keep children, because space is precious and we can't keep everything, the guitar or the CD? And the children answered, the guitar, the guitar. And that was the right answer, so they took the guitar and placed it in the bags with the other Last Things.

Then they rummaged around the ruins while the father and mother stood watch with their crossbows and the little boy found some old, rusted cans of peaches in the rubble of a house and the little girl found an old packet of dry seeds in an old tin. And the mother Guardian asked, So children, which shall we keep, the cans of food or the seeds? And the little boy answered, the cans, the cans! Because he was very hungry after so many weeks of scavenging and look-out duty in fear of the infected, and there were faded pictures of peaches on the front, and they made his belly rumble and his mouth dribble. But the little girl said, the seeds, Mother, because we cannot grow plants from cans of food, and besides they're dented, and they might be rusted inside or pierced and then we would get food poisoning and maybe die like Aunt Emma. And the mother and father Guardian praised her for her wisdom, and they discarded the cans and kept the seeds, and they put them in the collection of the last things, in the great bag that stretched to all dimensions, for it had many last things in it already.

Then the family of Guardians flew off in their spaceship made of junk to search for more of the last things, the things that needed to be saved, the things would be important to rescue and to pass on if humankind was to survive.

The Guardians of Last things was this story my dad recited to us since we were very small, a kind of interactive story, ever changing, because he'd ask us what we would we salvage and protect, from everything we could remember from our one day a week spent with him, TV sets and radios and tables and fridges, printers and paintings and mirrors, pens and pencils and roll on deodorants and A4 paper, and books and window blinds and chocolate and peanut butter and glasses. It was a way of thinking he secretly taught us, and we were never to tell any other parents or our mother or even our school friends about the Guardians. And every time we got to stay at his house, me and my little brother, in thenight, as we bundled down on his inflatable mattresses, he'd tell us the story again and ask us to choose one object over another, and to hold it tight in in our thoughts before we went to sleep. What would you choose – a tin opener or a knife? A computer cable or a rope? A battery charger or a candle? A gun or a spear? We can't carry everything, he said, so choosing is everything. And so, over those days and months, my brother and I secretly went through ever object we knew and worked out which to keep and which to discard, and we followed his instructions and hid these things in the bags he bought us, so that we would be ready for the coming collapse which he told us would be coming very soon. Whether it was the collapse of the modern world or his own mental collapse, I didn't know, until it was upon us.

ON FUCKEDNESS (3)

Where would New-Empathy leave our art? There was a time when nihilistic and confrontational art shook the lazy, complacent control systems of the period. 'Origin of the Species' dismissed notions of supernatural agency perpetuated by the Church. 'Naked Lunch' attacked McCarthyism and cultural repression. 'Last Exit to Brooklyn' reminded a conservative Western World that pimps and prostitutes had souls. Simone de Beauvoir had to write 'The Second Sex' to challenge biological essentialism. The safe baseline of the time necessitated such atmospheric disruption. Growing up in the 80's and 90's, it was cool to be nihilistic. It meant you didn't care in a world where stuffy, priapicauthority figures demanded you conform to their tenets of behaviour and taste. You were responding to an imperfect environment and there were still so many safe and generic control systems in place worth rebelling against. This world is different. We are stuck in that state of arrested development. We are teenagers with Alzheimer's in the Anthropocene[2]. The world is breaking apart. There is no denying the fact that technology offered us a retreat from this grim reality. We can let go of our collective responsibility, let the world die quietly around us while we gaze into the black mirror. Predictably, superficiality replaces self-reflection. It would probably be dangerous to be too self-aware in this landscape, or so those control systems of fuckedness would have you believe.

Distance and separation are no longer just tactics of the introvert, they are part of the default setting. Even as a dyed-in-the-wool introvert, I can see this is an acceptance of defeat, an inner-death. We have invited platinum-cured-informatic-qarînah, bogus 'machines of sunlight', into our bed chamber and there can be only one outcome.

Empathy underlies collaboration. Empathy is what motivates our engagement with the world and lets us see the reality of our situation – our world is dying. I'm referring not just to the ongoing destruction of our planet by climate change, but to the social fabric which gives our lives meaning. Now, more than ever, we should be looking to throw off the shackles of the Cave and directly tackle the problem we have right in front of our noses.

That's not to say we should be kind just to be transgressive. Our focus should be spiritual arcology (a fusion of ecology with architecture). The rebuild starts now and our art should reflect this because we are in the death throes. This is an attempt at a sensible, empathic, eschatological state of mind. When we need parody and satire we'll let you know. In the end times we do not need the violence of frustration. For us Westerners it is now more transgressive to be kind. In other words, it is now vital to express kindness.

Not phony kindness filmed for likes. Not commodified kindness, but real kindness.

Copyright © 2025 Chris Kelso and Ewan Morrison

Licensed to Merigold Independent
All rights reserved.

No portion of this book may be reproduced in any form without written permission from
the publisher or author, except as permitted by U.S. copyright law.

Cover Art by Cameron Austin of Ent Design
Interior Design by Saint Nick of Modern Alchemy

MERIGOLD INDEPENDENT

www.ingramcontent.com/pod-product-compliance
Lightning Source LLC
LaVergne TN
LVHW011632200926
848484LV00014B/2581/J
* 9 7 9 8 2 9 5 4 1 4 4 9 7 *